BLUFF YOUR WAY
ON
THE FLIGHT DECK

CAPTAIN KEN BEERE

D1077909

ℛℛ
RAVETTE PUBLISHING

Published by Ravette Publishing Limited
P.O. Box 296
Horsham
West Sussex RH13 8FH

Telephone: (01403) 711443
Fax: (01403) 711554

First printed 1992
Reprinted 1993
Updated 1995
Updated 1996
Reprinted 1997

Series Editor – Anne Tauté

Cover design – Jim Wire, Quantum
Printing & binding – Cox & Wyman Ltd.
Production – Oval Projects Ltd.

The Bluffer's Guides series is based
on an original idea by Peter Wolfe.

An Oval Project
for Ravette Publishing.

CONTENTS

INTRODUCTION

Shortly after it dawned upon the average hijacker that his anti-social pastime had less future than that of a chocolate oven-glove, many airlines returned to their convivial practice of allowing the occasional passenger to visit the flight deck. In truth, there never was much point in aping the American practice of locking the flight-deck door as no native British pilot could possibly survive for more than thirty minutes without a cup of tea, and all the prospective hijacker needed to do was to follow the stewardess to the flight deck with the milk and sugar and a well-oiled Luger. These days the pilots' greatest fear is that the tea is made with powdered Instant.

Visitors to the cockpit are a positive boon to air safety. On a longhaul flight, their imminent arrival can galvanize the captain into a frenzy of activity as he racks his brain to solve one down and complete the *Telegraph* crossword while simultaneously hauling his somnolent first officer from somewhere down by the rudder pedals.

It behoves a visitor, having created all this turmoil, to make sensible comments upon what is to be seen and to chat sympathetically about the tribulations of the average airline pilot. Here are two remarks which Jumbo crews take bets on.

'Isn't it small up here?' is the front runner, followed by, 'Don't we seem to be going slowly?' as the passenger peers through the front windscreens. At fifty pence a time, a shrewd captain can walk off with a fiver in beer money at the end of the trip if he chose the winner. To get attention, try 'Who owns this twenty-pound note on the floor?'

To bluff successfully in this subject it is necessary

not only to understand some of the technology involved but also to appreciate the sort of lives which the practitioners live. Psychologists put pilots' stress levels at the top of the league alongside surgeons, although the latter have an advantage – if the scalpel slips they don't accompany the patient to the mortuary. The pilot's stress is rarely apparent, except when the airline loses his suitcase.

Unlike the average Italian airline pilot who invariably takes off with the firm intention of strafing the beach at Anzio, the British version prefers to stay cool. Don't use all the bluff you will find in this book. Good pilots hold an honours degree in self-doubt and hate a smart-arse. Overdo it and the captain just might make you carry out the landing.

BEFORE THE FLIGHT

Much of the flight crew's work is complete before they ever set foot on the aircraft. Efficient planning is the basis of all successful flights. An hour before departure they check the flight plan, analyse weather forecasts, agree fuel loads, review aircraft serviceability and, most important of all, check if the pay-slips have arrived. In a large airline they may not even have met before so the initial work is accompanied by rapid mutual character assessment.

Flight Planning

To the great relief of the first officer, flight planning is rarely done by hand these days. A computer generated profile of the flight shows the route, headings, altitudes, recommended flying speeds, times between reporting points and fuel used, all calculated with respect to the forecast winds. Often uncannily correct, with the aircraft crossing each reporting point almost exactly to the minute shown on the plan, it can be wrecked by skittish isobars, or the refusal of **air traffic control** (ATC) to clear the aircraft up to its optimum altitude. Details of the proposed flight are filed with air traffic control.

For reasons of economy, airlines like their aircraft to fly with minimum fuel. For reasons of longevity, flight-deck crews like to fly with more fuel than this as they much prefer the engines to stop after landing rather than before. Negotiations before the flight often resemble an aviation version of the Hundred Years' War. In the end, the captain gets his way, just as his wife always says.

Fuel

Because of the extra weight, up to three percent of any extra fuel taken aboard will be used each hour simply to carry it. Thus, on a ten hour flight an extra thirty tonnes of aviation kerosene (paraffin if you are buying it at the corner shop) can use almost ten tonnes of itself simply by being there.

Fuel is measured in kilos rather than gallons because it is much more accurate. Like everything else, fuel expands and contracts with temperature. At your local Shell station, a gallon bought on a cold day will expand to more than a gallon if the weather turns warm, but it will still only carry the car the same distance. When filling up with thirty thousand gallons of the stuff it pays to be meticulous.

Weather

Intercontinental flights are planned to take advantage of favourable winds. It is often more sensible to cross the Atlantic with a following wind even though the route chosen is longer than one with the wind abeam. Flights within Europe are more proscribed and must follow the airways unless extreme weather causes an alteration to the route.

The weather forecast for the destination is a prime factor and a flight is always planned with a diversionary airfield in mind which anticipates good landing conditions. This second airfield is called, ungrammatically, the 'alternate'. The law states that enough fuel must be carried to get the aircraft to the alternate if weather at the destination closes in. So does anyone with half a brain.

Aircraft Serviceability

Nobody is perfect, and an aircraft is no different. If every single part were required to be flawless before it was permitted to get airborne it would soon simulate an expensively furnished garden shed. Loosely referred to as snags, the defects are not of the order to degrade aircraft safety and in a properly run company the captain always has the last word as to their acceptability. Snags are recorded in a book, the **technical log**. Copies are also retained on the ground, but it is much better not to try to deduce why.

Flight Notices

Aircrew are such transient creatures that the only way the airline can apparently communicate with them is via notices pinned to a convenient wall. The notices usually promulgate the news that the procedures promulgated last week are now considered dangerous and pilots should revert to the procedures from the week before which are now no longer dangerous but safe. Much of the subsequent flight is spent in argument.

Pre-flight Checks

Unless the crew-room coffee is too hot to drink, or the skipper's wife phones to remind him that the kids are growing up without him, the flight-deck crew get out to the aircraft before the passengers begin to board. Pre-flight checks are a vital part of the operation and cover virtually every aircraft system – electrics, hydraulics, pressurisation, flight controls, autopilots down to the windscreen wipers. Someone also has to do an outside check to make sure that all the bits of aeroplane are still there.

Often the captain does the 'walk round'. If the weather is particularly bad he delegates it to a grateful first officer. Should there be a flight engineer on the crew he invariably does it on the grounds that he is more likely to know how many engines the thing should be fitted with. And if the aircraft happens to be a Jumbo, it's a long walk for an older man.

Aircraft cockpit checks are completed either from a check-list or by following a laid down 'scan' to make sure that nothing is missed. If the equipment includes an **inertial navigation system** (INS) it is switched on early to allow the gyroscopes to spin up and the latitude and longitude of the aircraft position is inserted to tell it where it is. Get this wrong and you may as well invite a Bangkok taxi driver to navigate. You will often see the numbers painted high up on the wall of the **parking stand** where the pilots can see them.

It is during these checks that a member of the cabin staff brings in the first of innumerable cups of coffee or tea, most of which will go undrunk and have to be removed before take-off. This initial drink is invariably followed by the chief of the cabin staff who ritually grumbles about the airline, the flight, the catering, the ground staff, his roster, his pay and his family. Everyone agrees and feels better.

Bluffers should cultivate an expression of anguished sympathy for all members of the crew.

The Dispatcher

Essential to the atmosphere of general chaos is the dispatcher, charged with the risible task of getting the aircraft away on time. He it is who is responsible for co-ordinating the fuelling, the boarding of the passengers, and a hundred other things, any one of which can cause a delay for which he will ultimately be blamed. Many airlines require him to wear a distinctive hat which he has usually mislaid in the mayhem. He is still easily recognised by a harassed facial expression and mobile telephone swinging wildly from his belt like a loose stirrup as he gallops around the aircraft. To waylay him requires the agility of Will Carling. Nobody has ever seen an old dispatcher.

His final duty before sprinting off to cover the next departure is to present the captain with the **Loadsheet**. This legal document details the way in which the aeroplane has been loaded – passengers, luggage, freight and fuel. The information it contains determines the total weight of the aircraft at take-off and shows whether the aircraft is in trim. The aircraft cannot leave until the captain agrees the figures and signs it.

That the captain cannot possibly know whether the information is correct but must sign it anyway if he wishes to take off is one of those quirks of the system which he has learned to accept.

A good Queen's Council keeps this nugget of information so that if he is ever engaged in litigation of a flying nature he can get the captain weeping at the unfairness of life with his very first question.

Once the paperwork is out of the way, it is time to call ATC for a clearance and see if the engines will start.

GETTING STARTED

In an effort to disseminate the bad news as early as possible, aircraft are sometimes given a **slot time** by air traffic control well before their time of departure. This is the time at which the aeroplane must be airborne, and even at peak holiday periods it can often fall on the same day.

If the delay is lengthy, the decision must be made whether to keep the passengers in the departure hall and make them angry or put them on the aircraft and make them livid. Captains generally prefer to see their passengers safe on board, knowing from experience that to allow them to roam the airport for any length of time will result in many of them getting lost or taking the bus home.

Start-up Clearance

Even when just the normal chaos prevails, aircraft must ask for start-up clearance before winding up the engines. Sometimes it is given immediately or there may be an unexpected delay. If there is, it's very frustrating for all involved and some pilots can get quite ratty.

Of course, the fact that night follows day can get some pilots ratty, rattiness being their permanent state. Humour occasionally defuses the situation. After a British aircraft was given 30 minutes delay at a German airfield, a Lufthansa service was cleared to start immediately. The British pilot was furious. He demanded to know why Lufthansa was put ahead of him. Before ATC could answer, the Lufthansa captain broke in. "We got to the airport early," he said, "And put our towels on the runway."

Push Back

Most aircraft begin the journey in an undignified fashion by being shoved backwards with a truck. During this manoeuvre it is incumbent upon all concerned to ensure that there is nothing in the way. On the odd occasion when a member of ground staff unthinkingly replaces the pier and opens the aircraft door to retrieve his spectacles, the door gets knocked off and clatters to the tarmac. This is universally regarded as a bad start to the day.

If the aircraft succeeds in pushing back without taking part of the airport with it, it's time to start the engines. Why not start the engines before pushing back? Well, it makes for a quicker departure and avoids the embarrassment, in the case of a Jumbo, of the thrust from four large idling engines overcoming the truck and propelling it back into the terminal building.

Engine Start

Jet engines, in common with those in Grand Prix motor cars, are started by air. The primary source of this air is the **auxiliary power unit** (APU), a dinky little jet engine which lurks in the aircraft tail or lives tucked in close to the wing root. It's function is to supply electricity and air-conditioning while the aircraft is on the ground. When it decides to fail, it invariably waits until it gets to Bombay on Midsummer's Day. The APU itself can be started from the aircraft battery, often a surprisingly pathetic little thing which the average Russian wouldn't install in his Lada.

Quick deduction, for which all true bluffers are famous, will now dissuade you from joining the chorus of complaints about poor ventilation while fanning your face with the plastic emergency card and treating yourself to an unplanned nose-job. During start the engines need all the air they can get to spin up the turbines before the blue touch paper gets lit.

Engines are numbered from left to right when looking forward. Thus number four engine is the outer one on the right hand side (starboard) of a four-engined aircraft. This was logical when the engines were mounted forward of the pilot, but on modern aircraft they are behind him. The uninitiated might think that this could lead to confusion. It does.

Taxying

Modern airports have expanded to the size of small towns with their own, often quite complex, aircraft roadway systems. Many taxiways are one-way and others banned to large aircraft. It is not too difficult for an aircraft to enter the active runway by mistake. Unless the crew is very familiar with the airport's layout they resort to a map, the **airfield chart**. In the case of Heathrow the chart doesn't help much as the British Airports Authority (BAA) is always in the process of digging it up – summed up famously by one American skipper who described Heathrow as "The only building site in the world to have its own airport."

You'd think that with aircraft costing millions of dollars each they would fit a decent steering system. You'd be wrong. Just watch the average airliner weaving about the centreline as it taxies.

Steered by the nosewheel, the geometry is identical to that of an old Reliant Robin, though it isn't required to cope with the speeds that most three-wheelers achieve – the aircraft is normally airborne by then.

Wide-bodied aircraft are the trickiest to drive on the ground. It is almost impossible to see the wingtips from the flight deck of the Boeing 747 even though they extend almost 100 feet either side. Clouting the corner of a building with an aeroplane costing over 100 million dollars is generally frowned upon and can cost more than a captain's annual salary to repair. And if the pilot runs a wheel or two on to the grass it doesn't merely bump along like a garden roller. Supporting up to 400 tonnes of fuel and metal, the wheels generally bury themselves up to the axles, immobilising the aircraft. If the airport has to close as a result, the captain gets to see his transgression on *News at Ten*.

Before-Take-off Checks

These actions put the aircraft in the right configuration for take-off. Flaps are set to the take-off position and aircraft instruments checked, as are the trim and the speeds to be called on take-off. It is also quite important to get rid of the tea cups before the floor assumes its Alpine gradient on take-off.

In the cabin, the crew go through the emergency demonstration. No matter how many times you may have seen this, it is still a good idea to watch. Any stewardess will tell you that those who hide behind page three are the very first to start bleating for a repeat when something does happen. There is rarely time to do so, and positively no inclination.

Runways

Runways are defined by the first two digits of the magnetic compass heading upon which they lie. In aviation, the compass rose is divided into 360 degrees, numbered clockwise from north. East is thus 090 degrees and south 180 degrees. North is always called 360 degrees, not 000. If a runway lies east/west it will be known as runway 09 or runway 27 (just call it "Two Seven") depending upon which way you're looking along it. Helpful airport authorities paint the two numbers on each end of the runway and erect a signboard alongside the entrance. Most large international airports have sets of parallel runways, aircraft landing on one and taking off from the other. Logically, they carry the suffix "Left" or "Right".

Because the magnetic north pole wanders around the Arctic year by year, the magnetic compass heading of a runway can change. Thus, Heathrow's main runways which had been 28L (Two eight left) and 28R ever since the airport was built one day suddenly became 27L and 27R. To many older British pilots it was as though the Queen had announced that henceforth she would be known as Gladys and a few seriously considered retirement.

If there is no queue for take-off and the pilot is not required to pull into the holding bay to await his turn, the aircraft is cleared to enter the active runway.

TAKE-OFF

Even though the odds against it are calculated at hundreds of thousands to one, pilots always assume that an engine will fail on take-off. When it doesn't, they cheerfully press on with the flight. It's not that they are a naturally gloomy bunch, it's just that they don't like nasty surprises. Expecting the worst is a good system which has been proved over many years and reflects the fail-safe concept incorporated into all aircraft operations. Each take-off is based upon a couple of previously calculated speeds:

V1 (pronounced Vee One) is the speed on the runway up to which the pilots can forget the whole thing, slam on the brakes, select reverse thrust, stop safely and then go to the crew-room for a nice cup of tea. If take-off is abandoned above the V1 speed then crew and passengers will probably find themselves heading for the nearest Little Chef by the shortest route.

V2 is the lowest speed at which the aircraft can be flown with a failed engine. Engines being mounted on both sides, a failure of one causes the aircraft to yaw, something which must be counteracted immediately by firm use of the rudder. Below V2 speed there is insufficient airflow over the control surfaces to keep the aircraft straight. You do not get airborne before V2 – not more than once, anyway.

There is a nailbiting gap between these speeds on shorter runways which can present a ticklish predicament: engine failure when too fast to stop and too slow to get airborne. Then the aircraft must be kept straight on the runway using nosewheel steering and maximum rudder until V2 speed is achieved and the aeroplane can be safely lifted into the air.

Just so that you get the jargon correct, pilots insist upon saying that they have "lost an engine" as though they have committed an act of gross carelessness.

None of the above procedures apply to a single-engined aircraft where the traditional drill for an engine failure after becoming airborne is for the pilot to pierce the abrupt silence by calling to the passengers, "Say after me, our Father..."

Initial Climb

Having hauled the thing into the air one would be forgiven for thinking that the difficult part was over. The truth is quite the reverse.

Most airports publish a list of **standard instrument departures**, detailing routes and heights which must be strictly followed immediately after take-off. They are designed to separate incoming and outgoing aircraft and to avoid noise over the local MP's home. Some standard departures require manoeuvres which could win the world aerobatic finals. If the crew succeeds in jiggling the aeroplane through this aerial labyrinth without air traffic control demanding the captain's name, it's time to **clean up** the aircraft and settle down for the climb to cruising altitude.

'Cleaning up' involves raising the flaps and any other bits of aeroplane which are sticking out and spoiling the airflow. On smaller aircraft this generally entails whamming the flap lever to 'up', if it isn't already there, and getting on with normal life. On the Jumbo it is a more formal procedure carried out in stages and is as protracted and interesting as oiling a bicycle chain.

Objective number one then is to obtain ATC clearance to the requested altitude. Jet engines are greedy on fuel when low and a delayed climb can even mean not reaching the destination without a stop for more. Everyone is looking for height and another cup of tea.

Pressurisation

Aeroplanes are not "fully pressurised" as some would have you believe. To do so would require it to be pumped up to almost 15 pounds per square inch, close to normal atmospheric pressure, by which time the fuselage would resemble a tin rugby football. Nine psi is about all it is built to stand and even then the cabin becomes measurably longer – though it never appears to translate into extra legroom in the tourist cabin.

As altitude is gained, the atmosphere inside the cabin climbs at about 500 feet per minute until, at normal cruising, cabin altitude stabilises at a height of 8-9000 feet, a thousand or so feet higher than Mexico City but without the exhaust fumes.

If the pilots fail to pressurise or the automatic system doesn't work properly a warning hooter toots as the cabin altitude rises above normal level. If action is not taken immediately to arrest the rise in cabin altitude, down tumble all the oxygen masks, a sight worth seeing in any large aircraft as 400 devices burst from their housings to bounce merrily on the end of their plastic pipes. The scene is accurately described by the name commonly given to it – the rubber jungle.

Bluffers never stuff their ears with cotton wool for the climb or descent – it makes not a jot of difference to the discomfort. They simply swallow regularly on the way up, and pinch their noses and blow hard on the way down.

THE PILOTS

Don't expect to find a typical pilot. They come in all shapes and sizes and not one of them looks like Mel Gibson.

Who's Who on the Flight Deck

It helps to know who everyone is. The **captain** occupies the left hand seat and it's good sense to be nice to him if you feel inclined to be nice to anybody. He sports four rings on the sleeves of his uniform jacket and the epaulettes of his shirt if he has remembered to slide them on. If Sir appears to be four years old and is diligently sucking his thumb it is more than likely that the occupant of the seat is merely another visitor and the captain has just nipped back to the loo.

The **first officer**, second-in-command, slumps in the right-hand seat and is familiarly sometimes known as the co-pilot. Even more familiarly and for reasons not entirely clear, to the cabin staff he is known as "Nigel".

The co-pilot wears one, two or three rings to the captain's four, depending upon his seniority and experience. They betoken a second, first or senior first officer respectively. The latter rank denotes that he has the experience to become a captain but that there is currently no vacancy – at least, that is how the rank was first envisaged. Unless his captain is in the very act of buying him a drink in the bar of the night-stop hotel, the average senior first officer wishes him no harm but yearns for his abrupt demise.

Less frequently than of late you might find a third member of the flight-deck crew perched aft of the driving seats and gazing with varying degrees of incomprehensibility at a very large and complicated panel. This will be either another pilot or a **flight engineer**. If he shows any sign of understanding the winking lights and swinging needles, he is probably a flight engineer. Try lifting his briefcase. If it is apparently screwed to the floor then it is chock full of tools and he is most surely a flight engineer.

That's as many crew members as you will find in there. Never mention the rôle of 'navigator'. That his presence even crossed your mind marks you out as someone who probably last flew with Samuel Cody's Flying Circus. The navigator relinquished his job to a bundle of silicon chips years ago. The radio officer, with his clicking Morse key, is an even dimmer memory. Now the career of the flight engineer is on the same slippery path, making him a martyr to cost-cutting and integrated circuits. His final disappearance from the flight deck will be sorely felt. Apart from the loss of a crew member who knows which widget on the aeroplane is connected to what and can fix stewardesses' hair dryers, the greatest tragedy of all is that silicon chips are no help at all with the *Telegraph* crossword.

Fitness

Throughout his career, the pilot's ability and fitness is regularly dissected and analysed by other pilots and doctors who inspect him as avidly as the discoverers of the first coelocanth. Twice a year his competence and knowledge of emergency procedures is checked during a **Simulator Check**. Twice a year his body is probed for evidence of deterioration and debauchery in a **Medical Check**. Once a year his ability to operate a real aeroplane on a real route with real passengers is assessed on a **Route Check**. Throughout the year he lives with the frisson of knowing that any time a CAA pilot might leap aboard and plonk himself on the flight deck like a Bus Inspector intent upon catching out an errant driver. Sometimes irritated that his is the only profession singled out for such regular monitoring, he suspects that if the same competency tests were inflicted upon the medical profession, the crucial issue at the next general election would be over-population.

Licensed by the **Civil Aviation Authority** (CAA) and obliged to obey their regulations, he can find himself employed by an airline which encourages him to tweak the rules a little in the interest of the balance sheet. This, of course, he resists at some risk to his job and the expense of his frayed nerves.

As his primary source of information is usually the *Daily Telegraph* he may have a somewhat jaundiced view of the outside world. This innate conservatism is no bad thing: flying is firmly based upon rules and procedures and there is little room in the flight deck for those with alternative notions.

If married, it is best if he lives with a woman who is resigned to the life of a one-parent family and enjoys nothing more than coping with blocked drains and recalcitrant central heating. If unmarried, he is likely to be a martinet who makes all visitors to his home remove their shoes before stepping through the front door.

The one thing he shares with his colleagues, however, is an earnest desire to hold on to his job and avoid the appalling prospect of going out to work for a living.

Many pilots cut down their alcohol intake, adopt a strict exercise regime and begin to diet approximately one week before the date of their licence medical. The new lifestyle generally lasts as long as a politician's promise. There is nothing quite like passing an ECG check for opening up a vista of limitless debauchery until the next one is due.

Fitness is taken seriously, but it is invariably the health freaks who call in sick while the others grope, coughing, into the airport to fly the services. Never ask after a pilot's health – he shouldn't be there if he feels unwell and he is hardly likely to tell you if he is.

Simulators

The invention of the devil, simulators are an essential tool in the training and checking of pilots. Sitting in the pilot's seat and experiencing the surroundings, the sound effects and the apparent acceleration forces synthesised by the machine it is almost impossible to believe the whole affair is screwed to the floor. So sophisticated have they become that complete conversion courses are carried out in them, the pilots qualifying without ever flying the actual aeroplane.

Only those compelled to be tested within one can truly appreciate the full horror of the thing. If you can bluff your way into visiting one, accept with alacrity. It's better than Disneyland – if you are not a pilot.

When in conversation with any pilot, try to avoid using all words which remotely sound like "simulator".

Flight Time Limitations

Before the introduction of Flight Time Limitations, pilots were permitted to go 'crew fatigue' when they felt too tired to continue. They still can, but the option is rarely exercised. Calling an airline version of Robert Maxwell from Rome to tell him that you are about to book a couple of hundred rooms in the Hilton at his expense because you feel exhausted can lead inexorably to a short discussion of your career prospects. Legal limits to flying periods have helped to eliminate the need.

Flight time limitation regulations have resulted in airlines rostering their crews up to the maximum allowed as often as possible: pilots are now shagged out all the time instead of just occasionally.

Bidline

This can be the sole topic of conversation on a ten-hour flight. Within larger companies, pilots bid for a month's work line from a roster presented to them some weeks before. The more senior the pilot, the better chance he has of getting his chosen line. Those at the bottom of the seniority list end up with reserve lines and an intimate knowledge of the Third World.

Standby and Reserve

Required to go into the office on a working day, hang about for hours doing nothing in case you are needed to cover sickness or delays and if not, go home, sounds like heaven to most people. It's called an Airport Standby and pilots loathe it.

Reserve is equally unpopular. This entails staying at home for up to a month poised to dash to the airport within a couple of hours of being telephoned. When a pilot reflects upon what his colleagues are probably up to in Berlin, Bangkok or Barbados while he is hoovering the landing it can unhinge his mind.

Nightstops and Slips

To fly to another country, stay in a hotel for the night and return the following day is known as a nightstop. If you hand over the aircraft to another crew as you disembark then wait for a few days for the next service to arrive before continuing on your way, it's called a slip. Crews can get themselves into serious trouble on both, but have far more opportunity on a slip.

'Deadheading' is an American expression which the British call, less dramatically, 'positioning'. Crew schedulers regularly discover that the aeroplane is in one country and the crew in another. To the anguish of the accountants the two must be reunited by sending the crew as passengers, often on the airline of a competitor and paying the fare. The crews regularly get lost en-route by failing to hear announcements, visiting the duty-free shop or catching the wrong aeroplane.

Pilots make hopeless passengers.

Maxims

The airline industry is still too young to have many traditions. It borrowed its structure (**fleets**), ranks (**captain, first officer**), nomenclature (**forward, aft, port, starboard**), even its uniforms, from the Merchant Navy. Early aviators were taught by maxim, all of them invariably urging caution. Most are still relevant. A nod in the direction of an apt saying can allow entry to the club for a dedicated bluffer.

'There are old pilots and there are bold pilots, but there are no old bold pilots.'

'It's far better to be down here wishing you were up there than to be up there wishing you were down here.'

'Better to be a quarter-of-an-hour late in this world than a quarter of a century early in the next.'

Use the last one to a captain who has been under pressure from all and sundry to depart while he sensibly waits for the weather to improve and he will love you. It is a constant source of amazement to pilots that many passengers would much prefer to court death than accept a 30 minute delay while a blizzard clears the airport.

Who's Who among the Captains

The Management Pilot

Regarded with deep suspicion by all other pilots, the window cleaner, his wife and the dog. His decision to desert the flight-deck for a desk is considered conclusive evidence of a deep character flaw if not actual degeneracy. He emerges into the light to fly once in a while, but concentrates on the aircraft manuals rather than the *Telegraph* crossword. When met he often turns out to be disappointingly civilised and charming.

The Training Captain

Spends much of his working life in the gloom of a simulator training pilots who are undergoing a conversion course to a different aircraft type. Also tests pilots on their Simulator Checks. He can choose to sign a pilot's licence, or take it away. Sensible pilots are polite to him. Avoid bluffing this type – he has heard every variety.

The Route Check Captain

Sits on a spare seat in the flight deck trying to look inconspicuous while making notes about the crew's performance on their annual Route Check. He is not allowed to interfere. Is chosen for his patience and tact as any hint in his report that the flight was not a triumph of aviation expertise over prodigious odds is contested by all members of the crew as the ultimate travesty of justice and a case for the European Court.

The Line Captain

Has the best job in the world. Arrives at the airport, flies real people in a real aeroplane and then goes home or to the hotel at the end of the trip. His earnest wish is to be left alone to follow the profession he enjoys and, if his name be mentioned to the Management Pilot (above), he prays that the reply will be a puzzled 'Who?'

It is impossible to tell from the uniform into just which classification your captain falls. If he wears a mournful, haunted, expression he is probably a Management Pilot, but tread carefully.

THE CABIN STAFF

Many travellers are unaware that cabin staff are present on the aircraft primarily to deal with emergencies rather than minister to the passengers. Most of them fulfil both roles superbly.

After selection, the ground course is rigorous: by the time the crew are awarded their wings they have earned them the hard way. (NB: Pilots wear 'wings' whereas other members of the crew occasionally wear a single 'wing' which used to be known as a 'brevet' but is known illogically as 'wings' anyway.)

Dearer to the heart of airlines is the name badge which they are often obliged to wear. Never take these at face value: it is not uncommon for people to swap badges or invent fictitious names for themselves, so if your stewardess is called Arnold look around for some beefy steward named Clarissa.

Like the pilots, cabin staff undergo check flights and annual classroom checks on safety and emergencies. If they operate on several different aircraft types the tests can be a feat worthy of Mr. Memory of *The 39 Steps*. Many airlines have built mock fuselages to aid practice in emergencies and which can be filled with 'smoke' to imitate reality.

Longhaul flying demands a large share of equanimity. It is not uncommon for cabin staff to reach the penultimate day of a fortnight's trip and looking forward to getting home, only to receive a signal telling them to turn around and do the whole trip again. When this happens even a dedicated nomad would weep. More likely you will overhear the traditional comment – "You shouldn't have joined if you can't take a joke."

It is impossible to overestimate the bond of loyalty which can develop within a good crew. Woe betide those who offend a single member of it. If a passenger is obnoxious to one of them he is known to every other crew member within minutes of his transgression.

Cabin staff who are married to each other are often allowed the same roster so that they can be together 24 hours a day. Strangely, there is no recorded incident of homicide with this arrangement.

If your bluffing is not perfect, select the youngest cabin staff member to practise on. Experienced ones have heard everything and cannot be impressed. Flying a party of car salesmen to a conference an urbane voice was loudly heard to ask a rather beautiful chief stewardess, "Where have you been all my life?"

"Well," she retorted equally loudly, "I wasn't born for the first half of it."

After the initial excitement of the job, the ambition of most girls is to sleep, uninterrupted, for 24 hours. (If you are male and harbour unworthy thoughts you can see what you are up against – or rather, not.) The ambition is rarely realised. On many fleets, room parties are virtually obligatory. The disadvantage of being host is that you are inevitably the last one to bed and the first to apologise to the hotel manager for the previous night's uproar.

Cabin staff deal with hundreds of passengers each day of their working lives. They cope with people experiencing every emotion and exhibiting every foible known to man. When it comes to bluffing take this advice – try someone else.

THE SHARP END

Getting to the flight deck is often more straight-forward than is imagined and may be just a matter of asking. Plainly the first line of attack is through a member of the cabin crew and it helps if you can give a reason – never having seen the cockpit of that particular type, interested in the navigation, or simply pure enthusiasm. If you have a business card, send that up. It is the assurance of bona fides they're after.

If the captain or first officer visits the passenger cabin then things are easier. A request in person is rarely refused, especially when he can see for himself that you have just the one head.

Such opportunities as there are normally occur only during the relative quiet of the cruise. If there can be said to be a peaceful interlude while hurtling through the troposphere close to the speed of sound, seated in a thin metal tube surrounded by tons of inflammable liquid sloshing alongside barely contained blazing fires, then the cruise might well qualify for the description. The dedicated bluffer will take this view. Others are entitled to the opinion that, like the crew, he or she has a truly lamentable lack of imagination.

Even in earlier, less automated aircraft, the role of the crew during cruise is largely one of monitoring and talking to air traffic control. If the captain is wearing a headset and adopts a glazed expression just as you are about to pull your biggest bluff don't be offended – he is probably listening to a clearance from ATC. When his eyes refocus, just carry on where you left off – it's the normal mode of behaviour on a flight deck. If he removes the headset and his eyes remain glazed he is probably falling asleep and you should try one of the other *Bluffers Guides*.

Flight Decks

Flight decks changed very little until the last decade; designers just added more and more instruments and switches as aircraft systems became more sophisticated. They make for an impressive array but it is hell on a dark night if the lights go out.

The two **control columns** are interconnected, moving forward and backwards to lower and raise the aircraft nose. The control wheel makes the aircraft roll around its fore and aft axis – it's sometimes called 'banking'. Down below, the pilots' feet rest upon **rudder pedals** which operate, as you might expect, the rudder. Push the left rudder and the aircraft yaws to the left. If this is done without banking the aircraft to make what is called a 'balanced turn', every meal tray slides into the adjacent passenger's lap. Aircraft **wheelbrakes** are integral with the rudder pedals and can be operated independently, left and right, to steer the aircraft on the ground. Outboard on each side of the cockpit the small wheel is used for nosewheel steering.

More sophisticated aircraft, such as the poetically named European Airbus, are controlled by a small sidestick on the pilots' armrest which dispenses with the clutter and leaves more space for a meal tray. Allied with a 'fly-by-wire' system, this behaves like the kid next door – it gets an instruction and then decides if it feels like carrying it out.

Between the pilots is a central console upon which are mounted the **throttles** (or **thrust levers**, if you are American) which are pushed forward to accelerate the engines. Other levers on the console operate the **flaps** and **speedbrakes**.

There is plenty more ironmongery about, but these are the essentials.

In front of each pilot are the **flight instruments** – artificial horizon, compass, airspeed/Mach number indicator, altimeter, vertical speed indicator and turn and slip indicator. They are presented in varying degrees of complexity but all give the pilot enough information to keep the aircraft in the sky and flying in the right direction. Or at least let him check that the **autopilot** is still coping with the job. Both pilots have their own independent display.

The **airspeed indicator** is calibrated in knots (nautical miles per hour, slightly faster than mph) and works simply by measuring the pressure of the slipstream upon an external open-ended tube which faces forward – the **pitot head**. Once after landing with badly misreading ASI the pitot head was checked to discover a very angry and disorientated wasp glaring out of the hole.

As the aircraft climbs and the air gets thinner the airspeed indicator progressively underreads and bears less and less relevance to true speed. At altitude, the **Machmeter** becomes more important. This instrument shows the speed as a decimal fraction of the speed of sound and is often combined with the ASI. Most aircraft cruise at a speed between 0.8 to 0.86, the speed of sound being designated as 1.0 – if you see it indicating more than 1.0 you are in either Concorde or very serious trouble.

The **altimeter** displays height in feet. As it uses atmospheric pressure, it has a subscale with an adjustment knob to set the prevailing barometric pressure at local ground level to correct its reading. **Radio altimeters**, used for landing, are much more accurate and utilise a radio signal bounced off the ground.

The aptly named **artificial horizon** is equally important as it shows which way is up when in cloud or at night. In fact, pilots use it all the time and rarely consult the real horizon even when they can see it.

The **vertical speed indicator** (VSI) shows the rate of climb or descent in feet per minute. Expect to see it reading around 700 fpm descent on final approach.

Few of these instruments are now of the stand-alone type and are incorporated and integrated within displays which are difficult to interpret unless you did the pilots' conversion course. Understanding them quickly becomes second nature to the pilot. If he explains the intricacies, try to look interested.

The **instrument panel** between the pilots contains the engine indicators – N1 (revolutions per minute of the engine turbine), exhaust gas temperatures (EGT), engine pressure ratio (EPR – a direct measure of thrust) and other essentials. On a four-engined aircraft it can look complicated merely because the display contains four of everything. The layout is not accidental. When the engines are working normally the needles line up in a symmetrical pattern and any deviation is immediately obvious. This was particularly useful in the de Havilland Comet which produced so much power from its four engines that the pilots reckoned it was the only way they could tell that one had failed.

One thing you might recognise is the magnetic **standby compass** often mounted centrally up near the windscreens. It looks just like a car compass from Halfords and fulfils exactly the same purpose at a thousand times the cost. If the pilots consult this then you know the chips are well and truly down.

The Glass Cockpit

The glass cockpit cleaned up the general appearance of the flight deck but at first sight doesn't impress half as much. It is equipped with computer screens rather than individual instruments – screens capable of displaying more information than the pilot would need to build the aircraft from scrap.

Instead of the pilot scanning a large panel of individual instruments, almost all the information he needs to fly the aircraft efficiently is displayed on the screens in front of him. The design philosophy of the Boeing 747-400 ensures that neither of the two pilots ever needs to turn away from the front panel; all vital information, including failures, appear automatically on the screens and advise him of any action that's required.

The central one, the primary flight display, tells him just what the aircraft is doing – its speed, attitude, altitude, rate of climb or descent – the full list is enough to make your head ache.

The navigation display is less esoteric and much more interesting and is the one most proudly shown off to visitors. On this, a map of the route, radio beacons, the weather, nearby airfields, their runways and probably a moving picture of the chief training captain wagging his finger in admonishment, can be displayed after selection.

Above the central console, instead of the conventional engine instruments is another screen displaying a working picture of the conventional engine instruments. It is called the **EICAS** and not one pilot in ten can tell you what the letters stand for. Below it, the monitor writes messages to tell the pilots that say, number four engine has failed or, worse, that number three tea-urn has sprung a leak.

It's all very seductive, unless all the screens go blank and the pilots find themselves airborne in a small bare room. Happily, the aircraft designers insist there are so many backup systems that this cannot happen.

Recorders

Pilots are the object of more surveillance than John le Carré ever envisaged. Bluffers should know of the **cockpit voice recorder** (CVR) which faithfully stores the last 45 minutes of radio communications and flight deck conversation before recording over the top of it again. Inhibited by it when first installed, pilots have now reverted to their usual irreverent chat.

The second recorder is more sophisticated and called the **flight data recorder** (FDR). This gismo is an airborne detective agency, recording the actions of the pilots throughout the flight and the behaviour of the aircraft complete with any transgressions, which it logs for others to 'tut, tut' over at their leisure and then demand an explanation from the pilot who has by then forgotten that he even operated the flight.

The third device is the one beloved of journalists who insist upon calling it the **'black box' crash recorder**. Since its presence is redundant unless it can readily be found, it is normally brightly coloured.

Radio Communications

Civil aircraft normally talk to air traffic control by **radio telephony** (R/T). Fortuitously for British and American pilots the standard language used is English. Thus a German pilot talking to Berlin ATC will do so in that language and not his native tongue.

If you get the chance to listen in by being presented with a headset or by listening to the cockpit speakers, the first thing to ascertain is the aircraft callsign. Each airline has its own (British Airways – "Speedbird", or more often just the name of the operator "Air France" or "Lufthansa"). The name is followed by the flight number to complete the callsign – thus

"Speedbird 982" if you happen to be on the early morning BA flight from London to Berlin.

Aircraft within the same area of airspace all use the same radio channel, so the frequency can be extremely busy at times. Each hears all the calls from others within their vicinity together with the replies from ATC. It adds a measure of safety. Close to major airports the channels are more numerous and congested, and it can often be difficult for the pilot to get in with an essential call. In the USA the radio channels could easily be mistaken for a rap championship with all the finalists performing in unison.

There is one piece of vital radio equipment still used on some longhaul aircraft which cannot go unremarked. Policemen have their whistles, primitive tribes their tom-toms, and pilots have their **HF** (High Frequency radio). At the cutting edge of technology in World War II, this refugee from a ham radio enthusiast's shack with its static and fading reception is still in use for air traffic control in some parts of the world, though it is only really useful for picking up the latest cricket or football results on the BBC World Service.

It is not unknown for a frustrated captain to nip back and use the passenger satellite-telephone to call air traffic control. "Hello, you don't know me but I'm at present hurtling across the middle of the North Atlantic with 400 people and no clearance...Sorry, Mrs. O'Reilly, I was trying to call Shannon ATC...No, Patrick isn't with me right now...Yes, I know it's late but there's no call for that sort of language..."

The Phonetic Alphabet

Any bluffer worth his salt will know the current alphabet. For the record, however, it is: Alpha, Bravo,

Charlie, ('Coca' when first revised, but everyone was so sad to see 'Charlie' go that it was retained), Delta, Echo, Foxtrot, Golf, Hotel, India, Juliett, Kilo, Lima, Mike, November, Oscar, Papa, Quebec, Romeo, Sierra, Tango, Uniform, Victor, Whisky, X-ray, Yankee, Zulu.

Being international words they are easy to remember and pronounce in any language, but are not as emotive as the old 'Able, Baker, Charlie, Dog...'

Aircraft are generally known by the last two letters of the registration pronounced in the phonetic alphabet. Thus G-BOAC would be known as "Alpha Charlie" and never by the name written somewhere on the nose. Airlines are fond of naming their aircraft after Northern towns or rivers, convinced that passengers will then come flocking to fly on, say, the "Flower of Halifax". The name is usually placarded somewhere in the flight deck but most pilots avoid looking at it.

Aer Lingus appropriately named their aircraft after saints and used their incomparable Irish humour to christen the simulator 'St. Thetic'.

Air Traffic Control

Air traffic controllers have but one function: to stop aeroplanes banging into each other. This they do with remarkable success, especially considering the archaic equipment with which they are often provided.

On the rare occasions when things do go wrong, investigators spend many weeks trying to lay the blame on the pilots. If this fails they blame the air traffic controllers. It's a convenient system with a proven record of success and it ensures that airlines or governments don't need to spend money on expensive or non election-winning items like air safety.

Furthermore, pilots and controllers have come to expect it and would be unnerved if any other course of action were taken.

The quality of ATC varies but is generally excellent in developed countries. Amongst the very best is that in the UK. It is difficult to overestimate the feeling of reassurance engendered by the first words from a relaxed British controller when returning from a fortnight's flying around the world. The acknowledgment of his "Good Morning" can often come straight from the heart.

Turbulence

Nothing is more guaranteed to scare the pants off passengers than turbulence. A good bluffer keeps his cool – it's more a problem of comfort than safety.

The most irritating of the species is called **clear air turbulence** (CAT) when adjacent air masses moving at different speeds rub up against each other. At the boundary all hell breaks loose, air currents spiralling in all directions. As these areas can't be seen by eye or by radar and the Met officers can only make a despairing stab at where they might occur they are sometimes difficult to avoid. Usually the only course of action open to the crew is to request clearance to a different cruising altitude.

Research is in hand to develop equipment which will detect clear air turbulence before it spills tea all over the captain's lap. Ask about it. He is unlikely to know the latest but it will make you look good in the technical area.

Aircraft radar, however, can pick out thunderstorms. It's just as well, as they harbour the mothers and fathers of all turbulence. No sane pilot ever flies into a thundercloud. What about an insane one, you ask? Well, he'd have to be accompanied by an insane co-pilot and if that conjunction occurs then it's just not your day.

Standard practice is to allow the autopilot to fly the aircraft through turbulence – many autopilots have a turbulence mode for just this eventuality. But the only consistent thing about any type of turbulence is that it invariably strikes just as dinner is served.

Navigation

There is an essential difference between finding your way over the oceans and deserts or across a populated area like Europe. Navigating an aircraft to say, Spain, is often achieved by doing what used to be called a 'beacon crawl' – simply flying from one en-route radio beacon to another and 'homing' on to each one in turn. The ground transmitters are sited at intervals along the airway, making navigation relatively straightforward.

Most of these transmitters are Very High Frequency Omnidirectional Ranges, but to avoid running out of breath the pilots call them **VORs**. The pilot selects the bearing from the VOR which he wishes to maintain and follows the centreline until he is overhead. Usually paired with each VOR is a **DME** (distance measuring equipment) from which he can read off the distance to the station.

Still in use are some old **NDBs** (non-directional beacons) which are nothing more than a simple broadcasting station. Most airfields have one in line with each runway and about four miles distant.

Looking for a radio station in the middle of the Atlantic is, however, a fruitless occupation, and that's where the navigator and his sextant came into their own – until the introduction of:

The Inertial Navigation System

Navigation became a lost art the day **INS** was invented. As with many technological innovations in aviation, the principle was simple and the result phenomenal.

Mount three gyroscopes on a platform in three different axes. A gyroscope, like a teenage son on the sofa, resists any attempt to shift it. If the three gyroscopes are moved they push back and try to stay home. Measure how much and in what direction and you have the basis for a gismo which will tell you which way you are moving and how quickly. Naturally, it is important for somebody to tell the thing where on earth it is before it moves, because without that information it will never know where it got to. (This is the only time the navigator scored over the INS – he might have got lost, but he usually remembered where he'd started from.)

The INS's most precious attributes are that it requires no ground aids and doesn't care a jot about magnetic north, so that when over the north pole it is not confused by the dawning realisation that all directions are south.

When three Appeal Court judges deliberate, they do so in the knowledge that one might disagree with the other two but will be overruled in the final verdict. Aircraft, if they are to rely solely upon the INS for navigation, subscribe to the same philosophy and carry three independent sets. They differ only in that recent history shows Inertial Navigation Systems are far more likely to reach the right conclusion than Appeal Court judges.

Add a few more refinements and you have what is called, among other things, an **FMS** – a Flight Management System. Pilots love to demonstrate its ramifications and at the slightest show of interest will turn the tap on such an outpouring of information and calculations from the thing that the onlooker wonders why he asked. Mind you, when with ten hours still to go it announces that the aircraft will arrive in Singapore at 0103 GMT and the wheels touch the runway at exactly that time, it can be positively eerie.

At first installed only in longhaul aircraft, the Inertial Navigation System is now becoming standard equipment on shorter range aircraft.

When flying from say, London to Los Angeles, a city in a much more southerly latitude, don't ask why you can see Greenland out of the starboard window. Before leaving, take a piece of string and place it on a globe of the world between London and Los Angeles. The line along which the string lies is called the great circle track and is the shortest distance between two points on the earth's surface. Whenever possible, pilots fly great circle tracks because it is quicker, saves fuel and gets them to the nightstop hotel earlier.

Jet Streams

If you realise that the aircraft will arrive extremely early on your flight to Europe from the USA and have felt a little burbling turbulence during the cruise, ask if the aircraft has encountered a jet stream. These narrow bands of fast flowing air, travelling at up to 350 mph, are often found over the North Atlantic and can get you to Heathrow before the Underground opens.

BEHIND THE SHARP END

Wings

Bluffers should know about wings. After all, without
them the aircraft would be little different from an
overstaffed Number 8 bus.

In 1700 the Swiss family Bernoulli were blessed
with a son they named Daniel. He soon demonstrated
an alarming lack of interest in cuckoo clocks or army
knives and instead whiled away the hours investigat-
ing the motion of fluids. He discovered that when
fluid meets a restriction it speeds up to keep the flow
rate constant, causing a drop in pressure.

It became known as Bernoulli's principle, and air
behaves in a similar fashion. As it flows across the
top of the wing it speeds up to negotiate the built-in
hump, the pressure drops and generates a low
pressure area above the wing, making it rise.

Professionals are convinced the aircraft stays airborne purely by the will-power of the pilots who know
they will be in serious trouble with the management
if they lose concentration and the lift packs in. That's
why they always left one pilot on the flight deck while
the other drifted back to check if the gannets in first-
class had left any food. Now pilots are only allowed
out if about to wet their trousers and risk fusing the
aircraft electrics.

The trailing edge of the wing is hung with an array
of hinged panels which are called respectively
ailerons and flaps. Ailerons (French for 'little wings')
are normally the outboard ones and are connected to
the pilot's control wheel or sidestick. Their job is to
keep the wings level or to roll the aircraft into a turn.
If you watch carefully you can see that they are con-

stantly in motion, even when level in the cruise.

Inboard of these and often in several sections are the flaps. They are extended for take-off and landing, increasing the area and changing the shape of the wing to increase the lift and allow the aircraft to fly more slowly without stalling. Many aircraft also have leading edge flaps on the front edge of the wing. Once the aircraft speeds up all the flaps are retracted to restore the wing's basic profile.

Wings have a couple of further uses. They act as convenient fuel tanks and are an ideal place to hang the engines.

Winglets

These are the upturned wingtips you see on many of the later airliners. While useful in preventing preoccupied mechanics from falling off the end, they are really there to increase wing efficiency by preventing high pressure air from below nipping around the wingtip to the upper surface, ruining the lift and creating large wake vortices. Glider pilots, who need all the lift they can get if they are not to get home early, have used winglets since Pontius was a pilot.

Icing

Ice on the wings before take-off is not a matter for discussion any more. Modern high performance wings cannot cope with ice or frost contamination and are de-iced before take-off when necessary. If you happen to be flying with Cheapo Airways and see ice on the wings as you taxy out it's no time for bluffing. Get off – it's more than the aeroplane will.

Jet aircraft cruise too high for ice to be a serious

problem en-route. In the climb and descent wings are protected by hot air channelled inside the leading edge or occasionally by de-icer boots – rubber strips which can be inflated cyclically to break away the ice. Jet engines, also prone to icing, use hot air from their own plentiful supply.

Passenger Doors

It is one of aviation's great engineering mysteries that most aircraft doors, while opening outwards, are bigger than the doorway into which they fit. Designers call them plug doors and it took them years to work out how to achieve the impossible. Pressurisation, instead of pushing the door on to the heads of innocent bargain hunters in Hounslow High Street, merely holds it more firmly in place. It is akin to pressing the cork back into the neck of a bottle, a practice almost unknown to aircrew.

When the aircraft is fully pressurised, the air pressure inside the fuselage is around eight pounds psi higher than that outside. If the door is six feet high by four feet wide then there is a pressure of eight pounds pushing on each of its 3455 square inches. (Don't give up, we're getting there.) It all adds up to a total force of about twelve tons holding the door in its frame. Don't worry too much about little Willie opening the door in flight – a raging Arnold Swartzenegger would not succeed. Such fiddling with door handles simply results in a warning on the flight deck and makes for a certain amount of despondency as one of the pilots must get out of his seat to check the mechanical indicators on the door itself. And if it is the end of a long day, little Willie might just receive a well deserved clip over the ear with the captain's compliments.

CONVERSATION PIECES

Smooth bluffing requires that you have a ready topic of conversation available to fill those awkward moments when the captain realises that you haven't understood a single word of his description of the number three standby hydraulic system. Here are a few which are guaranteed to elicit a reaction.

Two-Crew Operation

Airline managers believe that nirvana will be achieved when pilots are dispensed with altogether – they are expensive, argumentative, and have very little respect for authority if it isn't wearing a pair of wings, and even then it is expected to buy the first round. Pilots also have a depressing habit of writing pertinent comments all over their orders and notices. Many managers dream of the day when they can shepherd their grandchildren to the Natural History Museum to view the last airline pilot, stuffed and mounted. Two-crew operation is the first step.

Like instant tea, its arrival on the scene was only a matter of time. Aircraft manufacturers simply built aeroplanes with only two useful seats at the sharp end and impressed prospective buyers with the cost savings. Automation made the concept possible.

Opinions within the pilot force vary but are predictable. Those flying aircraft with three crew aver that a two-crew operation is unsatisfactory, while two-crew operators announce that it is perfectly safe. You should count the crew complement before venturing to comment.

An odd feature of this commercial advance is that

because a two-man crew has a more restricted length of duty day than a three-man, it is not unusual to carry an extra crew member or two to share the enroute flying. Economics can sometimes strain credulity.

Heavy Crew

Extra pilots are known as 'heavy crew' and can be most useful to a bluffer. Examine your fellow passengers carefully. Heavy crew members often sit in the passenger cabin, but cheat by wearing a blazer or sweater instead of the uniform jacket. It would only be a bizarre pilot who would go to the lengths of changing his trousers, uniform shirt or black tie. Once identified he is your best ticket to the flight deck if you ask him nicely. He will probably say yes just to get some peace.

Crew Meals

Even crews have to eat occasionally. In the good old days the captain would stroll back to the first-class compartment to be served his meal in style. Only the wine would be omitted. Now he sits on the flight deck with a tray on his lap. To avoid universal food poisoning the crew members must eat different dishes. This is of little consequence to the diners as all the dishes taste the same.

It is not uncommon for a pilot to discover after relinquishing a tray that it was previously stacked on top of a colleague's meal, the remains of which now decorate his trousers, especially if earlier he was sharp with a stewardess.

A sensitive bluffer never says "Bon appetit" to a pilot unless they are both dining in a good restaurant.

ETOPS

This topic is a minefield, so try to avoid clumping over it too heavily. It stands for **extended range twin engined operations** and concerns setting off over large sheets of water like the Atlantic or Pacific Oceans in an aircraft equipped with only two motors.

There is nothing wrong with twin-engined aircraft *per se* otherwise civil airline pilots, who set great store by their pensions, would not dream of flying them. Twin-engined aircraft are just fine. Their only drawback is that when an engine fails they become single-engined aircraft.

This poses no problem over populated areas of the world. Look out of the window over Europe or the United States and you will spot more airfields than poppies in a Flanders cornfield. All around lie friendly strips of concrete just begging to welcome a slightly defective aircraft should the need arise. Protagonists argue that the fewer the engines the less the chance of engine-failure. (There is not a single recorded occurrence of engine-failure on a glider.) Multiple engine-failure often stems from a single cause which would stop sixteen engines if they were fitted. And statistically, the possibility of two engines failing for unrelated reasons are as infinitesimal as winning the National Lottery twice – though considerably less fun.

The protracted argument was finally clinched by the aircraft manufacturers building long-range twin-engined aircraft and demonstrating that they were cheaper to operate. Airlines love to save money. Just look at your meal-tray.

THE DESCENT

Pundits solemnly report that most aircraft accidents occur during the initial and final stages of the flight. As this is the time when the aeroplane is closest to the ground, thinking bluffers will already have worked this one out.

It is not the thought of accident, however, which causes the average pilot to sit closer to the edge of his seat when beginning the descent, but the fear of making a complete Horlicks of the whole thing and, say, ending up a thousand feet too high at the destination and having to buy the beer in the bar.

To save fuel, descents must be planned quite carefully. Of course, if the pilots are lucky enough to be flying an all-singing all-dancing glass cockpit affair it will work out the descent itself and do it while they ruminate on why they bothered to come along.

Ideally, the engines are reduced to idle power at the descent point and not adjusted until the aircraft is settled on its final approach path. Air traffic control try to wreck this plan by constant requests to stop the descent at intermediate flight levels or to fly at some impossible speed in the last direction which any pilot in his right mind would have chosen.

If the weather is fine, enthusiastic pilots often disengage the autopilot in the latter stages of the descent and fly the aircraft by hand. Not only is it enjoyable, but with the relentless increase in automation it is sometimes a comfort to discover they still can.

Speed Brakes

These are used in the air and do just what the name suggests: they stick out of the wings when selected and aim to slow down the aircraft in flight or increase the rate of descent. They also manage to create a praiseworthy amount of buffeting and pitching and can successfully unnerve passengers who are getting above themselves.

Ground Proximity Warning System

This prevents pilots flying into the ground. A legal requirement on British transport aircraft, it is not mandatory on many foreign carriers.

If the crew inadvertently descend too close to terra firma the **GPWS** calls a warning – typically, "Whoop! Whoop! Pull up!" giving them time to initiate an immediate climb out of danger. There is no discussion. They do it and talk about it later, after their bowels are once again under control.

Early GPWSs often gave the warning too late to take action and should really have been programmed to shout "Byeee!"

Holding

When there is a delay to landing, aircraft are directed into a nearby **holding stack** by ATC where they fly a racetrack pattern. Arrivals enter the stack at the top and everyone moves down 1000 feet at a time until the bottom level is reached and they are picked off by radar to start an approach.

Nobody enjoys holding, least of all the airline companies whose fuel is being used to no particular purpose, and watching all the other aircraft in the pattern soon palls. Occasionally one is joined by a Russian or Chinese aircraft flying a perfect pattern, but in the wrong direction.

If there is no question of a diversion, it can be a useful time to fill in a claim for more allowances or to solve twelve down and complete the crossword.

Diversions

In spite of the general passenger consensus that pilots divert in order to visit their latest girlfriend, crews hate to land anywhere but the original destination. It generates a sense of failure, makes for a lot of hard work and disrupts their social life (since they positioned their girlfriends at the original destination).

Aircraft rarely carry enough fuel to hang about for hours waiting for bad weather to improve. Eating into fuel reserves while flying around the holding pattern is an uncomfortable feeling for any pilot especially if he suspects that the weather at the diversion airfield might deteriorate. The decision as to how long to wait and where to divert is often a delicate balance of options. If you happen to be on the flight deck at this time it is best to maintain a low profile unless you want your head bitten off.

Intermediate Approach

Escape from the holding pattern is followed by the **intermediate approach**. The pilot gently slows the aircraft, extends the flaps and undercarriage and flies a heading which will eventually line up the aircraft with the runway. Whatever the weather it always demands total concentration from everyone on the flight deck.

At most major airports the aircraft will be directed through the intermediate stage by the ground based radar (Approach radar) and guided to a position from where the pilot can carry out his **final approach**. At an airfield not equipped with radar it is necessary to fly the **procedural approach** by following one of the **let-down charts** published for each runway. To stray off its route risks an argument with adjacent masts and mountains.

It can be a painful and time-consuming business often requiring the aircraft to fly overhead and then follow a pattern of tracks at designated heights to gain the final approach path. Some pilots would prefer to have a tooth pulled. With luck and good planning it will be the co-pilot's turn to do it.

Final Approach

About six miles from the runway threshold the aeroplane should be lined up with the runway and final descent begun. It looks fairly steep, but the optimum glideslope or angle of approach from the ground is only about three degrees – at the old Berlin Tempelhof airport the angle was more than four degrees to avoid the surrounding apartments and gave the impression of a helicopter landing.

With the flaps fully extended, the correct engine power set and a rate of descent of approximately 700 feet per minute the aircraft should be comfortably settled down without the need for too many alterations – pilots call it being 'in the slot'. It is generally accepted that a good approach leads to a good landing, although you can't always rely upon it.

If the wheels gently brush the ground and an increasing rumble from below persuades the crew they really have touched down no pilot can resist a sense of pride and achievement – especially if it is the right airfield.

THE LANDING

Getting there is pretty popular. If the aircraft doesn't land at the destination there was precious little point in leaving in the first place. Many millions have been spent in developing the technology to allow aircraft to land safely in all weathers.

Instrument Landing System

The **ILS** is the primary and most accurate radio aid used for approaches and is likely to remain so for many years to come. The "talkdown", much beloved by dramatists, is a thing of the past, and other methods of finding the end of the runway are relatively crude.

Aerials at the far end of the runway transmit two radio beams. The localiser beam marks the extended centreline of the runway, enabling the aircraft to follow it from many miles out and be certain that the runway is at the end. The glideslope aerial transmits a flat beam sloping up from the runway at an angle of about three degrees which defines the safe descent path.

However they are displayed on the instrument panel, the pilot's objective is to keep two indicator bars central – a vertical one which represents the centreline and a horizontal one which shows the glideslope. Any deviation must be corrected by an adjustment to the aircraft flightpath, left or right and by increasing or decreasing the rate of descent. The task can be undertaken by the pilot or given to the autopilot. In low visibility it is invariably handed over to the autopilot.

Once the aircraft reaches a certain altitude the captain must make an instant decision – land or apply the power and climb away.

Landing Limits

If you wish to look knowledgeable, ask the crew what **decision height** they plan to use for the approach. Do it during the cruise – they will be too busy later. Before you do, though, you had better know just what you are talking about.

Each runway has a minimum visibility promulgated for its associated approach aid. If the visibility is below this figure then it is illegal even to attempt an approach. Assuming that the visibility is above the minimum, the pilot may continue down to an altitude known as the decision height. When he reaches this height he must be able to see sufficient of the runway to carry out a successful landing – if he can't he must immediately forget the whole thing, apply power and execute a **go-around**.

This decision height depends upon numerous factors – aircraft type and equipment, ground aids, runway lighting and is typically between 300 feet and zero. Logically, the more modern and sophisticated the aircraft the lower are the landing limits.

Visibility is quoted by the control tower as Runway Visual Range (**RVR**). The old method of measuring it (to make out washing on a line five miles away) is plainly not accurate enough. What is required is the actual visibility along the runway.

The job was originally given to a fireman or other responsible person when the visibility was low. He would pass a miserable afternoon in the fog counting

53

runway lights and relaying the information to the control tower. Now the task is automated. Devices called transmissometers stand alongside the runway, like two skinny Martians facing up to each other for a duel.

Automatic Landings

It's a depressing thought to a pilot who prides himself on his skill and ability that he is allowed by law to land his aircraft as long as the visibility is more than about 600 metres, but has to relinquish the job to a collection of printed circuits if it is less.

In the present state of the art there is now no necessity to see anything in order to make a safe and successful automatic landing. The limit of around 100 metres visibility is imposed mainly to allow those at the sharp end to grope their way off the runway to the terminal building.

As with the INS, the usual complement of autopilots engaged in the operation is three. The concept is known as "fail operational".

Landing Lights

Early aircraft had little fitted in the way of headlights. Now most are equipped with more lights than the winner of the Alpine Rally. A pilot used to judge his height on landing by the perspective of the runway lights. Now he carries his own floodlighting.

Naturally, lights are also used for taxying at night, but also have a rôle in flight. Below 10,000 feet they are always illuminated on the principle of "see and be seen".

Touchdown

You won't believe this, but airlines like their pilots to make *firm* landings.

Aircraft tyres sit cosily retracted in the under-carriage bay until the time comes for landing – then some inconsiderate pilot on the flight deck drops them into a 250 mph gale. No sooner have they come to terms with the weather than they are dumped unceremoniously on to a hard runway.

The wheels are stationary. The runway is passing by at around 150 mph. The two have to meet up.

If the pilot does a smoothie, the tyres drag along the runway surface without spinning up which burns off rubber. A firm landing gets the wheels spinning with less loss of expensive tread. If the runway is very wet a good thump on to the ground also breaks through the surface water film and helps to avoid the dreaded hydroplaning. It's all very demotivating for a pilot, but provides the perfect excuse for a sudden dramatic arrival.

It's not uncommon for cabin staff to drop subtle hints after a heavy touchdown. One stewardess would stand disapprovingly at the cockpit door with her knickers around her ankles. Even the passengers assume they are qualified to join in. We recall an American who poked his head around the cockpit door as he disembarked and asked "Was that a normal landing or were we *shot* down?"

It is considered tactless to find this funny.

Crosswind Landings

Most pilots dislike these more than a Moscow night-stop. Aircraft are designed to land straight into wind.

Originally, of course, aerodromes were large open fields and the pilot could choose to land in any direction he wished. If the wind is strong and at an angle to the runway the pilot must lay off drift on the approach to maintain the centreline. In an extreme case he might find himself looking at the runway through the side window instead of the front windscreen. It's in this situation that he wishes he'd listened to his father's advice and joined the Civil Service, especially as the tricky bit is still to come.

Over the runway he is still flying sideways. Allowing the aircraft to touch down sideways is a surprise for the average wheel. As it is pointing towards the grass at the side of the runway any self respecting wheel will head off with its colleagues in that direction, taking the aircraft with it, and leading to much loss of face and seniority. The pilot must therefore straighten the aircraft with rudder before lowering it on to the concrete without undue delay. Any hesitation after straightening up allows the wind to carry it off the other side of the runway. It's not easy unless you are flying a Jumbo when the sheer inertia of the monster carries it down the centreline whatever you choose to do.

Reverse Thrust

Just after landing, as the passengers breathe a sigh of relief and inwardly sigh 'There, I knew we'd make it,' all hell breaks loose as the reverse thrust is selected, redirecting the engine jet efflux forward. Overhead lockers rattle, bags of duty-free slither along the carpet and the aircraft shows every sign of disassembling itself into a heap of alloy panels and rivets. Although reverse thrust can't compare with the brakes for slowing the aircraft, it helps reduce brake pad wear and overheating. And it helps prevent the aeroplane entering, for instance, the sewage farm at Heathrow, so it isn't all bad.

Unfortunately on a badly maintained runway it has a nasty habit of blowing forward stones and debris which can be ingested into the engine and do several thousand pounds worth of damage.

The defunct and unlamented Trident was unique in that reverse thrust was selected before the aircraft touched down and the aircraft could be lowered on to the runway by its judicious application. As this action was carried out by the co-pilot, it was the only air-craft ever designed to be landed by the person who wasn't flying the aeroplane.

Doors to Manual

Being able to explain this to a fellow passenger will raise your kudos no end.

Most modern aircraft are equipped with an inflat-able chute or slide built in to the structure by each door. It can be deployed to provide a quick means of escape when no steps are available – when, say, some-one notices a small ticking parcel in the next seat.

After the aircraft is taxiing under its own power the doors are placed into 'Automatic' mode by the cabin staff. This ensures that if a door is then opened from the inside the chute will automatically inflate and extend, offering all on board an exciting if inelegant slide to the ground.

As the aircraft approaches the pier after landing, the door controls are selected back to 'Manual' and the system is disarmed. If it is forgotten and the door is opened from the inside the automatically inflating slide will push the waiting ground staff back up the pier to the arrivals hall.

Parking

Forget landing in fog, forget severe turbulence, the trickiest bit of all for a pilot is parking the thing.

When approaching a pier it is vital that the aircraft stops within a few inches of the correct position. If it is even slightly misplaced the passenger door will not line up with the walkway and create great embarrassment as the ground crew meander off to look for a tractor to manoeuvre the aircraft into the right position while the standing passengers sit down again and grumble. Every airport uses its own peculiar system to guide the pilot and they all have just one thing in common – none of them work.

The aircraft nosewheel must follow the painted centreline and come to rest in exactly the right spot if the opening scenes of *Airplane* are not to be repeated and everyone enter the terminal through the plate glass windows. Some airports use traffic lights, others install angled boards which must be aligned by eye. Possibly the best system ever used was a ping-pong ball on a stick which hit the windscreen when the aircraft reached the parking position.

One final suggestion. As you disembark, inwardly congratulating yourself on a successful bluffing session, don't imitate your fellow passengers by thanking the the crew standing on the pier for a wonderful flight. They are ground staff and have probably just got out of bed.

HIERARCHY

Captain – Believes himself to be in sole charge of the aeroplane, an assumption secretly disputed by every other crew member and in particular by the chief steward. The captain's primary role is to sign anything put before him and act as duty scapegoat for anything which goes wrong on the aircraft.

First Officer – Second-in-command, does almost all of the work and gets almost none of the credit. His duties include feigning rapture as the captain describes in detail all 96 strokes of his last golf round or the difficulties encountered during installation of the new central heating system in his five-bedroomed house.

Third Pilot (if any) – Spends a depressing flight sitting sideways watching his panel and occasionally glaring at the backs of the other two having fun, until it is his turn to sit in the right hand seat and be a real pilot.

Flight Engineer (if any) – Flies with a healthy suspicion of the pilots' competence and the ever present possibility of their messing up his engines and airframe. A genuine mechanical wizard, he normally reports late at the airport because his car broke down on the motorway.

Cabin Service Director – Goes by many other titles depending upon the airline and is in charge of the cabin staff. On longhaul flights spends most of his time filling out forms and brewing himself endless cups of tea. If the cabin staff are a miserable bunch, go and inspect him and you might understand why. If they are fun, congratulate him.

Purser – Generally in charge of a specific cabin such as first class, business or tourist. May even carry a purse, although it is not obligatory.

Stewards and Stewardesses – Those who get to serve the food and drink, mop up the spills, check the loos, resuscitate passengers, restart hearts, deliver babies, deal with emergencies and take the blame for the airline's lack of investment in new equipment. In a perfect world most would be sanctified.

Ground Staff – Everyone else in the airline and invariably regarded by aircrew as guilty until proved innocent. Ground engineers and dispatchers are an exception to this generalisation and looked upon as honorary flying staff.

Crew Customs Officers – A unique and isolated class of individuals within the aviation hierarchy. Occasionally, a member displays characteristics reminiscent of the human race and is promptly transferred elsewhere.

VIPs – Royalty, pop-stars, politicians (of the ruling party), mass murderers and the Chairman's mother.

Passengers – The paying guests of the airline who can be guaranteed to: a) board festooned with more gear than Sir Ranulph Fiennes took to the North Pole; b) demand that morning's edition of the *Daily Express* in Bangkok; c) wait until the drinks trolleys are out before deciding to go to the loo.

NEVER ASK, NEVER SAY

"What's that town down there?" – Flying solely by reference to radio aids or inertial navigation the pilots don't have a clue unless it is Birmingham or Bombay. By the time the map has been unpacked and they have worked it out it will be fifty miles behind.

"How many gallons of petrol do we use on this flight?" – If it's a jet it uses kerosene and is measured in kilos and the poor old first-officer will have to work it all out.

"How long have you been a co-pilot?" – Longer than he considers humane and he may become angry or even begin to sob.

"What do you find to do when the autopilot's flying the aircraft?" – Work like one-armed paperhangers programming the thing and watch it full time to make sure that it is flying them in the right direction.

"Can you get *Eastenders* on that?" while pointing at the radar screen. Oh no, not again.

"You chaps must do well for duty-free." – They have much lower allowances than passengers. Most crew customs officers are evil to aircrew and would charge VAT on the haircut they had in New York if they could prove it.

"I was going to be a pilot but was failed because of colour blindness." Whether it's true or not, don't say it. Even those with the physical co-ordination of an airport trolley insist this was the reason. It's the aviation equivalent of growing too big for ballet.

plane – an abbreviation for "**aeroplane**," it is never used: "**aircraft**" is even better.

over and out – "**over**" means that the transmission is finished and a reply is expected: "**out**" means the transmission is finished and a reply is not expected.

portholes – although much aviation terminology is based upon ships, these are **windows**.

walls – partitions in the aircraft fuselage are known as **bulkheads**.

hostess – guaranteed to get up the nose of any **stewardess**.

circling – a journalistic expression usually applied to "**holding**".

coming in – unlike a rowing boat with its hour up, aircraft "**make an approach**".

air pocket – there never was any such thing and use of the word puts you last in the bluffer stakes. **Turbulence** is the word.

joystick – Biggles used one and made generations of schoolchildren giggle. It's called a **control column**. Or, if you are flying in an Atari derivative, a **sidestick**.

intercom – pilots talk to each other using this. Often used in mistake for "**cabin address**".

parachutes – there aren't any.

THE AUTHOR

Captain Ken Beere began his flying career with the Royal Air Force who taught him how to fly a jet fighter and the officially approved method of shining a pair of black square-bashing boots. Deciding that life was becoming too dangerous he bluffed his way into a career in the then unsophisticated civil aviation industry.

The RAF's gain proved to be BEA's loss, and he spent the next 32 years flying services in the highlands and islands of Scotland, then Europe, finally operating a British Airways Boeing 747 to countries of the former British Empire.

When not prostrate with jet-lag he wrote advertising copy for shaky companies, often administering the final poke which sent them toppling into bankruptcy. He renounced the pastime when companies offered to pay him good money to extol the virtues of their competitors.

He also edited a couple of aviation house magazines, one of which is now defunct and the other run by a damage limitation committee. In his editorial capacity, it is his proud boast that he never once rejected one of his own articles and often had to be restrained from sending himself a letter of congratulation.

Still a frequent flyer, he now agrees to undertake it only with a book by Garrison Keillor in one hand and a large malt whisky in the other while sprawled on a passenger seat in the most expensive class he can wangle.

THE BLUFFER'S GUIDES™

Available at £1.99*, £2.50 and £2.99:

Accountancy
Advertising
Archaeology
Astrology & Fortune Telling*
Ballet*
Bluffing
Champagne
Chess
The Classics
Computers
Consultancy
Cricket
Doctoring
Economics
The European Union
Finance
The Flight Deck
Golf
The Internet
Jazz
Law
Literature*
Management
Marketing
Modern Art

Music
The Occult*
Opera
Paris
Philosophy
Photography*
Poetry*
P.R.
Public Speaking
Publishing*
The Quantum Universe
The Races
The Rock Business
Rugby
Science
Secretaries
Seduction
Sex
Skiing
Small Business
Teaching
Theatre*
University
Whisky
Wine

All these books are available at your local bookshop or newsagent, or by post or telephone from: B.B.C.S., P.O.Box 941, Hull HU1 3VQ. (24 hour Telephone Credit Card Line: 01482 224626)

Please add the following for postage charges: UK (& BFPO) Orders: £1.00 for the first book & 50p for each additional book up to a maximum of £2.50; Overseas (& Eire) Orders: £2.00 for the first book, £1.00 for the second & 50p for each additional book.